Amazon Echo Dot:
15 Tips How to Operate an Echo Dot Like a Pro!

All photos used in this book, including the cover photo were made available under a Attribution-NonCommercial-ShareAlike 2.0 Generic and sourced from Flickr

Copyright 2016 by the publisher - All rights reserved.

This document is geared towards providing exact and reliable information in regards to the topic and issue covered. The publication is sold with the idea that the publisher is not required to render accounting, officially permitted, or otherwise, qualified services. If advice is necessary, legal or professional, a practiced individual in the profession should be ordered.

- From a Declaration of Principles which was accepted and approved equally by a Committee of the American Bar Association and a Committee of Publishers and Associations.

In no way is it legal to reproduce, duplicate, or transmit any part of this document in either electronic means or in printed format. Recording of this publication is strictly prohibited and any storage of this document is not allowed unless with written permission from the publisher. All rights reserved.

The information provided herein is stated to be truthful and consistent, in that any liability, in terms of inattention or otherwise, by any usage or abuse of any policies, processes, or directions contained within is the solitary and utter responsibility of the recipient reader. Under no circumstances will any legal responsibility or blame be held against the publisher for any reparation, damages, or monetary loss due to the information herein, either directly or indirectly.

Respective authors own all copyrights not held by the publisher.

The information herein is offered for informational purposes solely, and is universal as so. The presentation of the information is without contract or any type of guarantee assurance.

The trademarks that are used are without any consent, and the publication of the trademark is without permission or backing by the trademark owner. All trademarks and brands within this book are for clarifying purposes only and are the owned by the owners themselves, not affiliated with this document.

Table of content

Table of content 4

Introduction 5

Chapter 01: Amazon Echo Dot: An Amazing Voice Assistant 6

Chapter 02: Hacks to Get the Advantage of Alexa 13

Chapter 04: Hacks to Increase Your Productivity with Amazon Dot 27

Chapter 05: Hacks to Use Amazon Echo Dot as Your Personal Assistant 34

Conclusion 38

FREE Bonus Reminder 39

Introduction

Once and for all, the Amazon Echo has launched Echo Dot Second Generation. Almost everyone owning the Echo and companies are getting the Alexa support to the services and products by Amazon Dot. By using it, you are assuring that the excellent sound quality found in the houses' corner by the Amazon Echo Dot.

You will love it when you find that the second generation Echo Dot can be ordered directly from the Amazon, unlike the first Echo. Also, It has the better voice recognition, and it is very cost effective than the previous one. Echo Spatial Perception (ESP) helps to make it the part of your kitchen, hallway and a couple of rooms inside the house. In the scenario of multiple Echo, devices work together, but the only nearest voice issuing command will answer.

After reading this book, you will be able to understand that how to connect the Echo dot? The Amazon echo is the powerful home sound system at your home. It is your all time buddy and will never let you alone.

Chapter 01: Amazon Echo Dot: An Amazing Voice Assistant

Amazon intended to provide its virtual voice assistant into all over the home.

The second generation, Amazon Echo Dot smart speaker is not only mini-sized; it is the best in hearing, too. It also can connect via already setup at home. However the best-in-all smart home speaker is only Amazon Echo Dot.

Basics put-together and use

The basis put together of Echo is very easy. It is as simple that you operate your mobile. It does not require any training which can be bothering for many of the people. One echo device can cover an area of 1000 to 2000 square yards.

The thing that needs consideration for the better performance of Alexa is that you put it in the room where you spend the time mostly. All you need to do is:

Plug in the echo

Echo has its cord and needs to be connected to the power supply. Connect that cord with any of the switch from where Alexa can hear your voice.

Connect with App

The use of Alexa requires you to connect to the internet. To connect to the internet, it is necessary that you download its app. The app is available on the internet, and you can download it easily. It is free of cost. As you download the app, you can get the connectivity of Alexa with the internet. But make sure that you are connected to the internet as well.

Use Alexa

Now as you get the access to the internet as well to the app of Alexa, now you can easily enjoy the services by Alexa without nay nuisance. You can listen to the song, news, weather forecast and much more.

 It can also provide the information regarding any address, any place an anything of which you can think about. It is very user-friendly, and you will love living around with it.

Simple to Set Up & Use

1. Plug in Echo
2. Connect to the internet with the Alexa App
3. Just ask for music, weather, news, and more

Parts of Alexa

Alexa consists of several parts which work together in synchrony to provide the vast variety of functions. These are small parts and do a large work. The list and working of the parts will be discussed here:

Dimensions

Alexa is 9.25 inches in height and is 3.27 inches wide.

Volume ring

It has a volume ring. You can use the volume ring to modulate the voice according to need. You can move the ring in a clockwise direction to increase the volume. You can move the ring in a counter clockwise direction to decrees the volume. The white color of the light ring will tell you the intensity of volume.

Woofer

It has a woofer that is of 2.5 inches. The woofer is of high quality. It is made to deliver the deeper base with clear sound.

Reflex port

A reflex port is also incorporated in the Alexa. The function of this reflex port is to give clarity to the sound. It has the ability to deliver deeper sound with clarity.

Tweeter

The Alexa has its tweeter which is 2.0 inches in size. It is used to deliver the high notes of the sound and is effective in giving the right tone.

Light ring

One of the unique features of Alexa is its light ring. The light ring shows various colors that are intended to show the activity on which Alexa is working at that moment. Some of the colors that Alexa's light ring shows are:

Blue with cyan light

This color shows that Alexa is starting up.

No light

It means that the device is all set to take your commands

Solid blue with cyan color towards a person

This indicate that Alexa is busy in taking the request from the person where the cyan is facing its point

Orange light moving in clockwise

This indicated that Alexa is connecting to your Wi-Fi connection.

Red light

This color indicates that your microphone is off. This gives you the indication to turn on the microphone for working of Alexa

White light

White light appears when you are adjusting sound of Alexa

Violet light in oscillation

This indicated that there is an error with the connection of your Wi-Fi

Action button

The action button is the essential part of any device. It has following functions:

- It wakes up Alexa
- It can be used to turn off the alarm when you set one
- It can also provide you access to the Wi-Fi set up when you keep it pressed it up for 5 seconds

Microphone off button

The purpose of this button is to turn off the microphone. When the microphone is off, the ring shows red color. Press this button to turn on the microphone.

Power LED

The power LED shows the status of connection of Alexa with the Wi-Fi.

Power adapter

The Alexa has its power adapter that allows the user to connect it with the source of power.

7 array of microphone

The Alexa is loaded with 7 array of a microphone which allows coverage for the vast area.

Remote control

You can also buy a remote control to operate Alexa. The remote control is available separately. This remote allows you to operate Alexa even if the microphone is off.

amazon echo

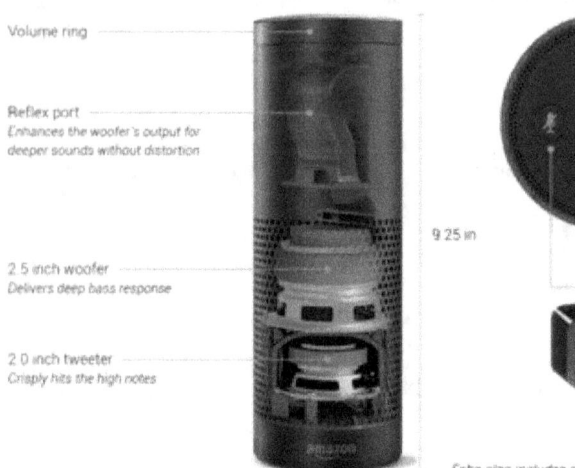

Chapter 02: Hacks to Get the Advantage of Alexa

Alexa is a virtual voice turn-on assistant. It is connected by Amazon cloud. Alexa has Siri in speaker, which is a female voice. By calling her name you just wake her up. Siri can not only wake up on her calling name, she also alerts on "Amazon" or "Echo." The Echo Dot has inside microphones that is always listening and when they prompt for the word "wake up" they just record every word that is said and send that word through the cloud to Amazon's server. When it happens to those servers, they just figure out what is Siri has been asked to do. Then the servers tell Alexa how to answer and this conversation process takes only about a second.

1. What is for next- kitchen help, set timer, and Alexa can keep you updated by reading you a custom news while you are having a morning refreshing breakfast

Alexa makes your whole of the kitchen smarter. Make dishes with the Alexa to help you remind the recipes that you selected for dinner time, and it keeps time preferences. **Alexa** can keep you aware of **USA today's** news and you do not need to miss your breakfast for reading the newspaper. Here Alexa reads official news for you when you ask her for News Today. You can also hear **the Buzz Feed News Briefing.**

When you are cooking, alarms and timers are big helps to make you alert when you are busy in kneading dough and your hands are covered in it. Not in position to touch your phone with dough covered fingres, so **Alexa, set the timer for 10 minutes"**, and then you know Alexa will notify you immediately.

You forget but Alexa do not. You empty the last bit of your cat food. Just say loud to Alexa to add in the Amazon shopping cart. So **Alexa, reorder the cat food.** In fact you can order AA batteries or the paper towels. You are being able to add anything from last shopping list. This will help you as the kids will do more than your knowing, they cannot sneak and order Xbox.

2. **Alexa gives you hearing: the radio, a podcast, Discover Weekly on the Spot**

   Here, some shared and daily commands:

   - When you command that **"Alexa, Go and search my favorite NPR"**, it plays local NPR station.

   - When you are in the mood of your favorite local station you just command that "**Alexa, play** [your favorite local station]**.**"

- Many times you are interested in sport and want to hear form sports area then you just command that "**Alexa, it's my sport time, you need to play something interesting for me. Go and find Sports Radio.**"

- Having fun mode, so just command some comedy station and say that "**Alexa, Its time to Pandora, play the best comedy of the time for me now.**"

- Do you love to hear podcasts, so do not skip your podcast serials, you just have to make a wish like that "**Alexa, as I love to know about my serial podcast, you will also like it. Find my favorite serial podcast of BBC.**"

- Randomly select your favorites in one play list so that you can easily enjoy your songs with one command that is "**Alexa, get my one of the best show Discover Weekly and move it on Spotify, I am waiting.**"

-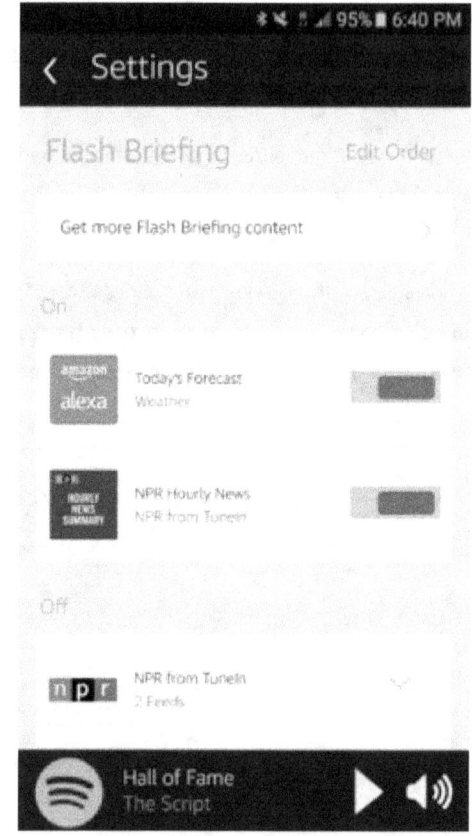

3. **You can access a version of the Alexa app online at <u>alexa.amazon.com</u>.**

   It gives the history options and setting from any App, or you can view all settings and history.

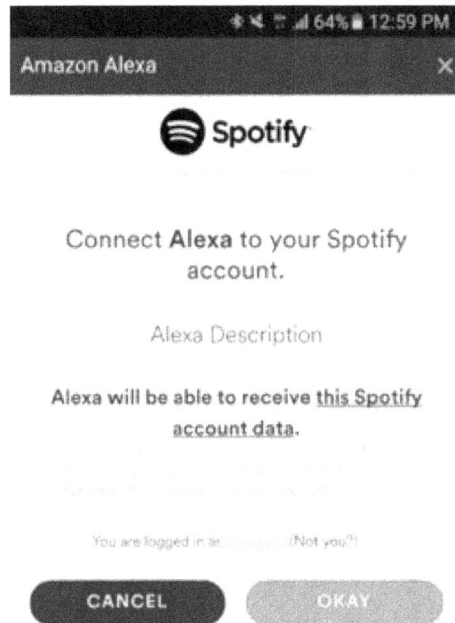

4. **Read, Tell, Ask, and Repeat: Alexa,**

   Alexa read my book.

   Once Echo connected to user's audible account, when you use book read commands for reading any book, it allows the user to hear book reading through speakers. During the book reading, users are allowed to skip, resume, and more. Now Kindle library books can add in to Amazon Echo. It can read non-fictional books where tone can be ignored. The famous speaker David Sedaris reads "Me talk pretty one day" in an effective tone, so Echo might not be capable of getting such fictional tone but it can read as well other media literature, too.

Alexa, tell a Democrat/Republican joke.

You want to hear something good and just change the mood, so you command the Alexa to tell you some political jokes, she will and definitely she will tell the joke, but not sure it can be good one or the one that you hear before.

Fun with Alexa

Laugh time! Say something funny to Alexa, and she will not disappoint you at that moment. She gives you a smile, and you can laugh and find Alexa fart without any smell. Oh really she can- it farts, for fun only. **Alexa, ask for a fart.** No, we have not told our kids about this one yet.

You want that Alexa copy you. So you need only want to hear about ALexa can repeat whatever you say. Instantly you command, "Alexa, Sarah says..." it is more fun with kids, use Echo's remote control from another room, and when they are playing and suddenly hear you "kids, take your shoes off" it will always be surprising moment for them and their face are like astonished. How Alexa can command them instead of their mother.

Chapter 03: Hacks for Media and Music with Amazon Dot

5. **Voice training: Alexa's talkback is the voice train of your accent. The Echo keeps the user voice recordings, and administering the recording that can catch on the app. You can, however, delete the recordings from the app.**

   The Echo keeps built-in seven microphones. All the microphones will alert on the word "Alexa" in the regular talks. Does it make sense? But in times past you will feel the Alexa is your best talking partner except it, Alexa can do a lot for you. Then it will make your mind and life happy.

   Alexa should understand the way you talk and recognize your sound. Alexa will serve you more when you speak loud and clear to it. Alexa recognition capabilities can also improved when it opens the App Alexa, it can access on both iOS and Android. You can follow these steps, first open the Alexa App then go and hit the Menu, you can select **Settings** inside the Menu, then you will find your desired option of **Voice Training**. In usual it takes few seconds (you have choice for different phrases say out and loud), during the training of Alexa, you are capable to stop it at any time.

   One question arises in mind that Echo can share among the user and other members that are only guests or part time roommate. In this case who has access to control the Alexa App. And user can authorize the other member, too. They can command it by using their own mobile phone and Alexa can hear all the requests. You cannot be ignoring all the time that "Echo keeps both ear", because Echo hears all and access user can listen them on mobile. So make sure that you are turning the voice off.

You want to clean unwanted recordings so just perform few steps and delete all the unnecessary recordings. Alexa History, that keeps voice record. You can delete as well your voice recordings by simply tap the Menu and then select the delete Voice card. Further, you want to rid of your all voice history so you have to delete your entire voice recording. Another option you will get when you are interested to delete whole history. You need to select the option **Devices** which gives you more inside of your Echo. Then you will go to select the option Manage **voice recordings, and the last option for all history that you can wipe out is named as Delete.**

6. **The ultimate Voice command list of routine**

Your wonder time is over by providing the simple and amazing ways to talk to Alexa. We are performing routine tasks and it is simple to ask or tell different things to Alexa.

7. **The smart home appliances connected with and can control via Echo.**

People believe that magic comes with Alexa. You can spread the magic of voice controlled smart home, only under the 50dollars. Now how do you turn your home into the smart home? The smart plug can do it. So you need the only smart plug to get connected as the smart home. People prefer few of the smart plugs like

Wemo switch plug with the cost of 40dollars and the TP-Link the smart plug that costs 35dollars. Smartly, you can get in practical plug any appliance in this plug known as curling iron, electric blanket, fans. All you need to say, **Alexa, turn on my humidifier.** Another command is **Alexa, turn on my bedroom light.** Wink home hub also a plug and use it for the smart purpose. Give the name to the device that is plugged in and use command to turn off and on for it. Like **Alexa, turn down the lights.** Phillips Hue Smart Bulbs can be coonected via the Echo and all those bulbs can control through voice commands. In short,you can say that Echo controls lights, switches, and thermostats with compatible Samsung Smart Things, Philips Hue, Nest, WeMo, Wink, Insteon, and ecobee smart home devices

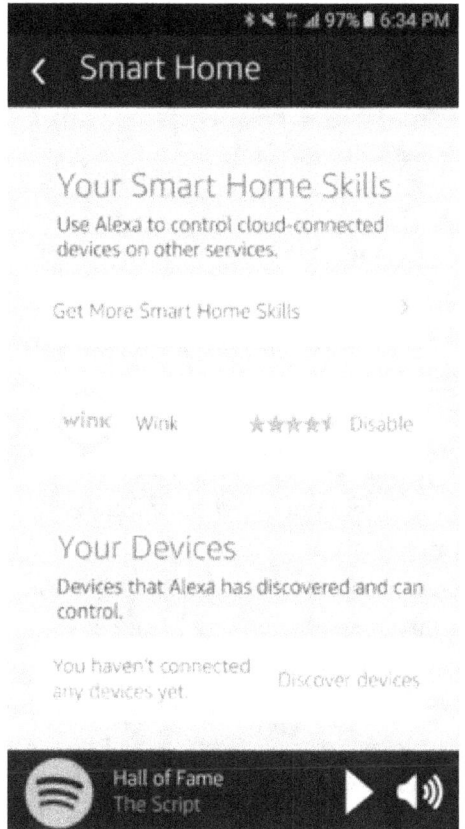

8. **Your phone and computer can control the Spotify at Echo . You chose the default music player is Spotify. Wondering how two different Spotify accounts can play music with Amazon Household.**

You do not need to be bothered about the necessary steps to connect your Spotify.

First You can find the Alexa app, then you go fo the Alexa **Menu**, there you will select the option of **Music & Books**. And finally you can select your own choice of Books as well Music.

You might want to create your best songs in a playlist or make you best music collection playlist.

"Alexa, Find on Spotify and then play my all time favorite songs playlist."

Never disappoint by leaving your jazz party songs in the middle just Alexa helps you to stay tune from where you made your last hearing, **"Alexa, Get in Spotify** this will enable you to continue to listen from the place where you left. This will help to swing your mood, too. Want some great moments then ask **"Alexa, Hope you would not mind and select Classical Jazz and play my Spotify in this moment."** Queue songs feature you to relax and feel free to set next song for listening. Control and Control, you do not need to be all time with your footsteps towards the audio setup because Alexa can checked from multiple sources where it has connected. You love when you will be able to control the Echo by Spotify desktop or mobile App. Like Echo volume can increase and decrease by phone or computer. Same as you can Pause or Skip the unwanted Lyrics. Remotely access your Echo is interesting as it's another feature.

If you do not select your default music player then the Echo will just put up Prime Music that is a smaller library than the Spotify. For Echo, help is always ready. The Amazon Alexa App will lead you to the Menu, then you will have the **Settings** option and you can find also the "Account" option then you can select **Music Media**. Now you can search and further go to the **Customize option where you can customize you albums and music. then the option you will select is " Choose your default services**. You can select you default music services here. And the last step is under the "Choose default music services," select **Spotify**. Another option to get connects and selects the default

Spotify, (in case of against the default selection, open you internet web browsers and then hit the alexa.amazon.com , and you will find it on the website.)

In preference, the youngest ones who have roommate can also detect Weekly playlist. It only required profiles of the multiple users. Go and create it. In the Alexa app, go to the option **Settings**, then find another option under the settings is Account, then select it and tap **Profile** you will have more options of profile type, so select Household profile and then last step is **Add**. Now you add your household profile in Alexa App.

Wonderful twist mode is the feature to provide different user profile. When you want to switch accounts, only say "Alexa, I am not quite sure you are on my account, So you are ready to Switch account" or "Alexa, whose account are you operating now." Different users can add their accounts in the Alexa App.

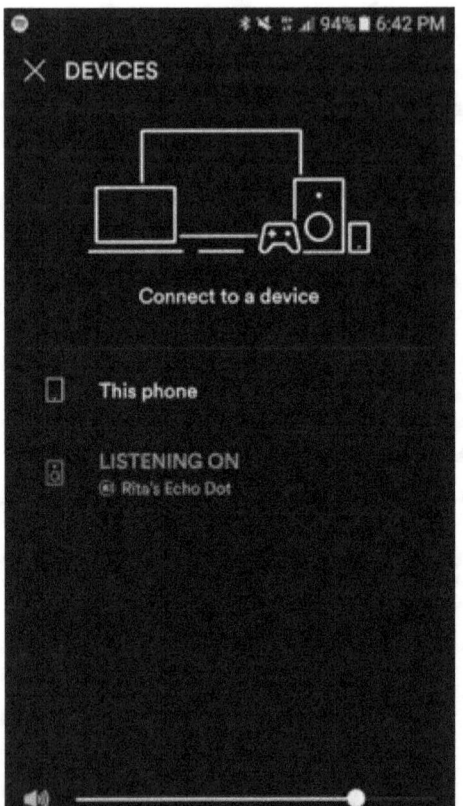

Chapter 04: Hacks to Increase Your Productivity with Amazon Dot

9. **The Echo can turn into the Bluetooth speaker as well.** For Amazon Music, the Amazon Echo equipped with the hands-free voice control. When you still think that you have lost something finding like it could be your podcast and your personal collection. Problem solved, by using your Echo as a speaker and both devices connected via Bluetooth. Then it is also known as enabling the Bluetooth pairing mode. Then only need to say "Alexa, do not wait and get pair." It will connect and your Echo becomes your mobile phone speakers. When you finished, and you do not want to continue any more pairing session then say, "Alexa, You are done now, and let disconnect."

The Voice-Operated speakers are incredible when Alexa turns into its Dj, and the dance party begins when you just say aloud and name any artist, or song, or music. For more fun and loud sound the Dot lets the audio setup lovers a brand new power. It can just connect with external speakers as well by using its Bluetooth. Still, Echo likes to rocks with the beats.

10. The amazing Echo "skills" apps than any other. Kids love her because the Alexa is a buddy-buddy in the games and entertainment, and a Translator, too.

Alexa's skills are one in the downloadable Classic Gaming references reorganization. It can bring up the many crafting techniques in Minecraft. And notify you that who will be the Destiny's upcoming armor weekly sale.

The games are exciting too, '**Alexa, start the animal Game**' will let you assume an animal then Alexa will ask a series of Yes/No questions and reveal about the guessed animal. **Alexa's Easter Eggs** is popular for the kids' interest. **Alexa, Start Movie Quotes',** to match and find the great dialogue where that originated.

Alexa Combat is entertaining, too. When you just want have fun with your sword, then say *Alexa, open combat and hit* the dragon with my sword. And it will fall you to excite in any sort of conflict with entertaining results.

No more thinking for understanding the language, because Alexa know you better than others. So it has feature to translate language in your desired one. You need it and Alexa, **ask Translator to how to say "please clean your room" in Italian.** Alexa is a capable of fifty-languages. She can translate it for you. Alexa translate it and find it on the Alexa App. Unfortunately, multiple foreign language skills that possessed Alexa are not the repeatable. So she translates and back to its native language in a minute. So you quiet have patience and ask again for the next translation.

Alexa, a calorie counter, You are a night binge eating lover or a diet conscious, you just always worried about the calorie counters and think is this high in calorie or low in calorie. When you fail to count calories and make it Alexa job to find the calories in your ideal food and meal. She will make sure how many calories are you taking by sending commands to its server and find the required response. You might want to know that how many alories in a pizza slice, so need only to shout that..Alexa, find calories in a pizza slice.

11. IFTTT added features to Alexa, An informer, build a Habit

Here, it is vital to explain about IFTTT that Different software and gadgets connected through IFTTTT and the great thing is that it has a channel entirely for the **Amazon Alexa.**

Echo can connect with the multiple devices and software such that your smart phone, EverNote, it is known as the custom "recipe". Another option for Echo is a number or existing recipes. It prompts to find your phone; it can also change the light color when playing song changes. EverNote Checklist can update by adding the list of pending jobs.

Alexa supposed to be your informer, when you do not watch the current news regarding politics, sports and showbiz then Alexa is here to give you all about it. Alexa will look for the info which you curiously want to know about, like sports schedule and scores, and who won the Oscar or the year when Princes Diana answer. She answers Math's problems, too. You can never underestimate the Alexa in logical problems. Alexa is not less than any competent tutor. She solves math and helps you in your budget.

More people believe on sayings and they truly rely on its originality. The more difficult is to change something although it could be buying a car, or trying new cuisine, or following a diet plan regularly, or trying to lose weight, and so on. But everything goes with one similarity that is changing habits. Like for car you need to change you buying habits, for weight lose you just have to make less diet for yourself and regular go for thirty-minute cardio or some physical fitness program, for balancing your diabetes, you have to regular monitor sugar level and avoid sugar in your food and kitchen, too. But saying is any habit changes and build in twenty-one days. Developing new business and personal habits you are supposed to covering a journey, where the **Alexa, start 21 days** will help you to the guide during your developing habit building plan.

12. Device is behaving not well: Finally, reset the Echo

Only try first the plug-off and re-plug-in the device. If it does not work, then reset Echo the Factory Setting. A reset button is located beneath of the Echo. You just have a breath and grab the smaller button that is called reset button you can hold it for at least ten seconds and the light wrapped in a ring turned to blue, and then it turned to orange and got in the setup mode. It indicated that the Echo has successfully reset.

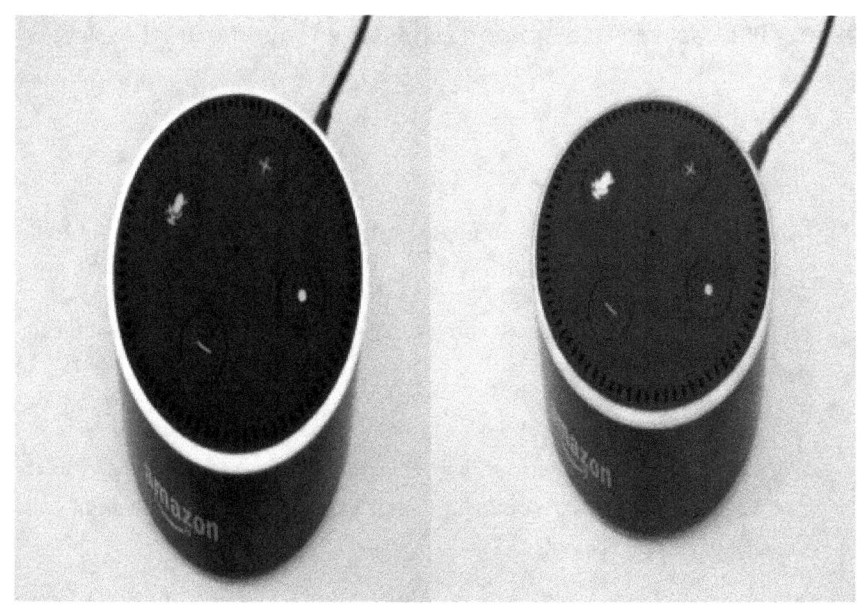

Chapter 05: Hacks to Use Amazon Echo Dot as Your Personal Assistant

13. **Alexa is a great to-do list manager. Add Google Calendar and let Echo tally the dates and calculations.**

    You can ask and get the Alexa's correct response for *"How many days until December 25th"*.

    Command to "**Alexa, What's on my calendar today?**" Alexa will search to find the relevant event. App and it remind your day up activities while you are just having a sip of your morning coffee. It is awesome, to remember all great number of jobs that parents had to do in a day.

    - No more worries to switch between multiple apps now you just one step away to turn the Echo into your own personal assistant! Reach the Alexa app > Settings > Account > Calendar and link to your Google account.

    - Alexa helps to manage things like "Alexa, what's on my calendar this weekend?", "Alexa, what is new on my calendar on Sunday?", or "Alexa, when is the wedding?"

    - Alexa can organize your shopping list as well, you say to her that Alexa, add shampoo to my shopping list. Alexa will not just add into your shopping wish list it also confirm you by saying "YES."

Get latest price by **Alexa, Bitcoin Price and find the Bitcoin Price Checker,** you are aware of the return price of Bitcoin in US dollar. Alexa will tell you the exchange rate of the currency by checking coindesk.com. Information provided by Alexa, Open the

Bitcoin Rate. It will give you the current rates of 1BTC in the required base currency included useful market data.

All time work and responsibilities give you stress, Alexa is here to make you happy. She makes your quality time and she gives you smiles by telling the jokes. When you ask Alexa, Bubble Boy

14. **By adding work address location in Alexa, you will be able to ask Alexa about your daily commute, and APRS. Alexa Automatic- a problem solver, too.**

    Alexa helps you to manage your daily commute by telling the roads to avoid in the morning. It keeps you away from the traffic rush, and it saves your time by reporting early about the traffic. How to activate Traffic prompt in Alexa: In the Alexa app, go to Settings > under Account, open Traffic > add your home and work addresses.'

    No more guide for finding the location of APRS, Alexa, use APRS and locate [kilo victor six mike dash seven], say "Alexa, Locate APRS station".

    You are facing daily hacks of auto, where I parked my car, how much gas as left in my car, or how much kilometers I have driven, "Alexa ask Automatics, where my car is?"

    You need an automatic adapter to avail the Alexa Automatic skill.

You only want to know BART timings. What is BART advisory? Well, no more worry to get the current BART advisories. Or simply say Alexa, tell the BART status.

Next you can also get the trains schedule, Alexa is here for you. Alexa, open BART Times. It gives you the alternate trains between you house and destination station. And also tell you about the live service advisories and information of Elevator status.

15. Find nearby local businesses and nearby food outlets by Alexa. Also Alexa can also Find BART schedule

Alexa, open "Skill," it is a trigger word that is required by Alexa, and it is needed to say when you have wished to call some skills made by another company like Uber, Lyft, Domino's Pizza, Fitbit and more.

How you can ask things like "**Alexa, what Chinese restaurants are nearby?**" then Alexa responds and, you can ask follow-ups: "**What is the phone number?**" "**How far is it?**" or "**Are they open?**"

- "**Alexa, find the phone number for [name of restaurant].**"
- "**Alexa, find the hours for a nearby beauty salon.** "

Alexa's other skills are Alexa keeps you relax, "**Alexa, one minute sound meditation of mindfulness**", it enters you into the peace world for one minute. Alexa makes you fit, ten-minute workout in the assistance of Alexa, Get your cardio and reduce your stress by asking **Alexa, start a ten-minute workout.**

Alexa, Battery Boot, hey Alexa which one battery boot is the best for you. Echo battery boot buying store can suggest from Alexa. So Alexa, **ask battery boot where to buy it?**

Alexa is wonderful to family group reward system. It tracks points like for virtual bean jar.

Many of the Echo's fans love the Alexa boxing skills. Alexa, Launch beat cylinder. Alexa offers her beat-boxing skills. Simply you ask for a beat by name such that "old school" "phat" "electronic" and etc.

< Lyft   < Domino's Pizza

  Lyft   **Domino's Pizza**
Lyft   Domino's Pizza, LLC
★ ★ ★ ★ ★ 10   ★ ★ ★ ★ ★ 29

| Enable Skill | | Enable Skill |

Account linking required   Account linking required

"Alexa, ask Lyft for a ride."   "Alexa, open Domino's"

"Alexa, ask Lyft how much a Lyft Plus from   "Alexa, open Domino's and place my Easy
home to work costs?"   Order"

"Alexa, tell Lyft to rate my driver five stars."   "Alexa, ask Domino's to track my order"

Conclusion

It is easy to understand that users can accomplish a lot about the Amazon Echo Dot and Alexa when they will purchase it and use it. The Second-Generation, small -sized is potent than its first effect. With the significant hearing capabilities, it is available at exceptional prices. What is unique, an existed audio setup can connect via the only Echo Product.

This book effectively helped to set up and use your Echo. Voice Controlled Alexa Echo Device gives the freedom. For Amazon Music, it caters hands-free voice control. Ask for all artists and songs, or want a particular genre. Also, it enables users to search music by Lyrics, Albums or songs bring in the market, or have Alexa's choice of music for you.

Here, it is evident to mention that Echo Dot is officially available in the US, UK, and Germany. So, if you are using it another region of the world, very few features does not work as expected. In the end, I would like to thanks the reader to download this book.

FREE Bonus Reminder

If you have not grabbed it yet, please go ahead and download your special bonus E book *"Chakras for Beginners. 7 Steps To Understand And Balance Chakras, Radiate Energy, And Strengthen Aura"*.

Simply Click the Button Below

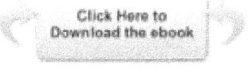

OR Go to This Page

http://lifehacksworld.com/free

BONUS #2: More Free & Discounted Books & Products

Do you want to receive more Free/Discounted Books or Products?

We have a mailing list where we send out our new Books or Products when they go free or with a discount on Amazon. Click on the link below to sign up for Free & Discount Book & Product Promotions.

=> **Sign Up for Free & Discount Book & Product Promotions** <=

OR Go to this URL

http://zbit.ly/1WBb1Ek

www.ingramcontent.com/pod-product-compliance
Lightning Source LLC
Chambersburg PA
CBHW050030230526
45470CB00003B/1212